YOUR KNOWLEDGE HAS VALUE

- We will publish your bachelor's and master's thesis, essays and papers

- Your own eBook and book - sold worldwide in all relevant shops

- Earn money with each sale

Upload your text at www.GRIN.com
and publish for free

Bibliographic information published by the German National Library:

The German National Library lists this publication in the National Bibliography; detailed bibliographic data are available on the Internet at http://dnb.dnb.de .

This book is copyright material and must not be copied, reproduced, transferred, distributed, leased, licensed or publicly performed or used in any way except as specifically permitted in writing by the publishers, as allowed under the terms and conditions under which it was purchased or as strictly permitted by applicable copyright law. Any unauthorized distribution or use of this text may be a direct infringement of the author s and publisher s rights and those responsible may be liable in law accordingly.

Imprint:

Copyright © 2016 GRIN Verlag, Open Publishing GmbH
Print and binding: Books on Demand GmbH, Norderstedt Germany
ISBN: 9783668297050

This book at GRIN:

http://www.grin.com/en/e-book/340058/analysis-of-the-scientific-contributions-of-wilhelm-conrad-roentgen-and

Mark Zaidi

Analysis of the Scientific Contributions of Wilhelm Conrad Röntgen and Marie Curie

GRIN Publishing

GRIN - Your knowledge has value

Since its foundation in 1998, GRIN has specialized in publishing academic texts by students, college teachers and other academics as e-book and printed book. The website www.grin.com is an ideal platform for presenting term papers, final papers, scientific essays, dissertations and specialist books.

Visit us on the internet:

http://www.grin.com/

http://www.facebook.com/grincom

http://www.twitter.com/grin_com

Analysis of Scientific Contributions of Wilhelm Conrad Röntgen and Marie Curie

Syed Mark Zaidi

2016/09/12

Contents

Introduction 3

The Impact of Gender Inequalities in Scientific Revelations 4

The Role of Chance and the Scientific Method 5

The Risks in Pursuing Answers in a Newly Discovered Field 7

Conclusion 8

References 9

Introduction

Throughout centuries, mankind has always been fascinated with our innate surroundings. Every moment of every day, new revelations in the sciences persist to clash with our understanding of the world we live in, threatening to toss and turn us to an immeasurable extent. Rather than being fearful of these new revelations, scientists throughout history take it upon themselves to explore broad new frontiers, in hopes to better comprehend the fabric of nature. Paraphrasing Tommy Lee Jones, from the hit movie *Men in Black*, "Fifteen hundred years ago everybody knew the Earth was the center of the universe. Five hundred years ago, everybody knew the Earth was flat... Imagine what you'll know tomorrow." (*Jones, 1997*). What one may know to be true now does not guarantee it to be the truth, only the true understanding that we are able to attain. Several hundreds of years ago, people would have cackled at the notion that there is a light that cannot be seen by the naked eye, yet can pass through a block of wood with ease. Furthermore, people would have chortled at the thought of a female scientist being able to make any noteworthy contribution to the fields of chemistry and physics, let alone discover not one, but two new elements. Wilhelm Conrad Röntgen and Marie Curie stood boldly in the face of current beliefs and understanding, solely to progress society's understanding of the world we live in.

The purpose of this essay is to help one comprehend how it is that the two had been successful in the aforementioned task. It is no easy feat to move the entire world forward in its current understanding of what we live in. Many adversities and obstacles had to be overcome, such as the thought that men were superior to women, back in the 1800's. Additionally, it is prudent to investigate how the element of chance had played a role in the discoveries pertaining to Curie and Röntgen. The scientific method we know and love today does not always guarantee the best results. Sometimes, sheer luck can accomplish what may take decades, if not centuries. It is also vital to acknowledge that despite taking the necessary measures to ensure your health and wellbeing, there is no guarantee that the experiments you may perform will not harm you. Science is dynamic, not static, and understanding the methods behind great scientists will allow future generations to progress the world forwards through to a new era.

The Impact of Gender Inequalities in Scientific Revelations

Wilhelm Conrad Röntgen had lived a life with relative ease, being a successful professor of physics, and winning a noble prize for the discovery of X-rays. He worked in a state-of-the-art laboratory, and had access to presumably whatever he had needed to further his research. Prior to working on X-rays, Röntgen had sufficient time to publish 49 different scientific papers (*Röntgen, 1896, 343*). However, relative to Marie Curie, Röntgen had lived a life of ease. As one could tell from her name, Curie is female. And back in the 1800's, even up into the late 1900's, it is a well known fact that females had not been treated as equals to males, as it should have been. Unlike Röntgen, Curie did not have a lavish laboratory. Rather, she had been given a small room in the School of Physics, described by Eve Curie in *The Discovery of Radium* as "a kind of storeroom, sweating with damp, where unused machines and lumber were put away" (*Curie, 1943, 368*). Not only did this take a physical and mental toll on Curie, this environment would have likely delayed her research by years, as opposed to having a laboratory similar to that of Röntgen. Having proper equipment would have likely led to a faster discovery of the elements Radium and Polonium. Furthermore, it was likely the case that Curie had not earned the same amount of money on a daily basis as her husband, let alone the majority of other males in a similar social class. A lack of funding and resources can significantly delay any research project. Stuck with low grade equipment, materials, Curie was put up for a task that could have been accomplished in a fraction of the time, had she possessed suitable gear. Though, despite the many hardships she had to endure, Curie was persistent. 45 months after announcing the probable presence of Radium, Marie Curie, along with her husband Pierre Curie, the Curies had proven its existence (*Curie, 1943, 379*). However, it pains me to think of how much time could have been cut back, had she been treated as an equal. Time that could have been spent finding new discoveries was taken away, solely due to having an 'X' chromosome in place of a 'Y'. Seeing as Curie had lived only to the age of 66, time truly was of the essence (*"Marie Curie Biography", n.d.*). However, it is pleasant to note that at least Pierre treated her with the respect she so rightfully deserved, and this is also analogous to another case between Antoine and Marie-Anne Lavoisier, which was one of the few cases where females in the history of science were treated as equals to males. Marie and Pierre Curie had banded together, united by a passion just like Antoine and Marie-Anne Lavoisier, and brought the world through to a new era.

The Role of Chance and the Scientific Method

It is jocose to think that if it were not for Röntgen being a tad disordered, none of the aforementioned and following discoveries would be possible, and the state of scientific understanding would not be what it is today. Of all the materials that could have been on his laboratory bench the day Röntgen was recreating an experiment by William Crookes, what are the odds that there would be a sample of potassium platinocyanide, let alone any fluorescent material? Typically, an object would fluoresce when bombarded with light of a specific wavelength. In Röntgen's laboratory, the sample of potassium platinocyanide was being exposed to X-ray radiation, which would cause the sample to fluoresce. I've personally attempted to recreate a variant of this, but instead of using toxic potassium platinocyanide and harmful X-rays, I had used a safe uraninite-enriched glass marble, along with ultraviolet light (a neighbouring wavelength to X-ray, but much safer) to demonstrate the effect of fluorescence, as noted in the following *Figure 1*:

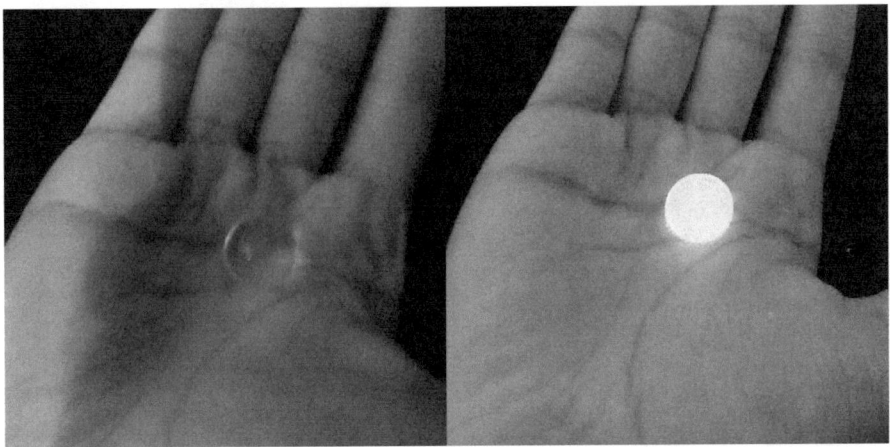

Figure 1: Comparing a sample of uraninite glass under normal lighting (left) to ultraviolet lighting (right). © Syed Mark Zaidi, 2016.

As noted from the above juxtaposed photographs, the uraninite-enriched marble glows a brilliant green. Without such a bright light being emitted, Röntgen would have never seen the potassium platinocyanide fluoresce, thus prompting him to ponder up the existence of X-rays. This supports the notion that the scientific method is not always the best method to yield results. That is not to say that the scientific method (observe, hypothesize, and test) is not beneficial. Shortly after the potassium platinocyanide fluorescing incident, Röntgen began to experiment with a wide variety of substances, ranging from planks of wood to lead glass. Prior to the experimentation, he had not known that density played a role in limiting the propagation of X-rays in a medium. After finding out that denser substances limited X-ray propagation in a medium, he began to test this hypothesis with materials of a known density, such as platinum, lead, zinc, and aluminium. I have attempted to demonstrate this phenomena using an inert gas discharge tube that emits high-frequency waves that radiate through a low-density sheet of cardboard and fluoresce the phosphor coating on the inside of a fluorescent lightbulb, as noted in *Figure 2*:

Figure 2: A fluorescent lightbulb glowing due to radio waves being emitted from an inert gas discharge tube placed behind a 4 millimeter thick black cardboard sheet. © Syed Mark Zaidi, 2016.

This experiment differs from that of Röntgen's setup, in the sense that I used a much slightly shorter wavelength for safety reasons, and a fluorescent lightbulb as opposed to a photosensitive film. Nonetheless, it conveys the concept that shorter wavelengths of electromagnetic energy are able to pass through low-density materials.

Amusingly, if it were not for the chance ensued upon Röntgen's experiment, Curie would have likely never performed her feat. This is because Henri Becquerel would have never attempted to test uranium's fluorescing properties, only to find that it emits its own radiation, and thereby, prompting the Curies to delve into the mystery behind radiation. In more lament terms, if the potassium platinocyanide was never left on the laboratory bench, Röntgen would have not discovered X-ray. Therefore, the Curies would have never begun their awe-inspiring work with radiation. It truly is amazing how a case of mild disorganization can influence the advancement of the sciences by such a gargantuan extent.

The Risks in Pursuing Answers in a Newly Discovered Field

Marie Curie had lived to the age of 66 (*"Marie Curie Biography"*, n.d.). Röntgen lived to the age of 77 (*"Wilhelm Conrad Röntgen - Biographical."*, n.d.). While it is true that the average life expectancy back then was a lot shorter than what it is now, there is no denying that the work the two were involved in shaved a considerable number of years off of their life. With the amount of radiation that the two were exposed to, it is no surprise that Röntgen died from carcinoma (*"Wilhelm Conrad Röntgen - Biographical."*, n.d.) and Curie died of aplastic anemia (*"Marie Curie Biography"*, n.d.), which were likely caused by the huge amounts of radiation that they were exposed to. It truly is a shame, considering that nowadays, there are many simple innovations that can protect oneself from harmful radiation, such as lead-lined barriers. Even I had taken some precautions while performing the aforementioned demonstrations in *Figure* 1 and *Figure* 2. Though, one may ponder that why had they not considered using such precautions? Simply put, no one was aware of the harm. Since X-rays and radiation were emerging fields, there had been no knowledge of the harm that they can cause to organic tissues. And it did not help that the two were exposed to it on a daily basis. Röntgen was obsessed with X-raying everything, including his own hands, while Curie was in direct contact with radioactive substances on a daily basis for nearly 45 months (*Curie, 1943, 379*). And since the effects of radiation poisoning are not quite immediate, no one had thought that they could harm anyone. This can even be applied to current situations. What we may believe to be safe and harmless can just as easily be killing us slowly from the inside out. Analogous to the aforementioned quote by Tommy Lee Jones, in the early 1900s no one had thought that smoking cigarettes would have been harmful, but in less than a century, the current scientific understanding at the time proved otherwise, as it is well known that cigarettes can cause many diseases, such as cancers, among others. What we believe to be a safe practice now might not be the case in a century or so from now, as scientific knowledge is always dynamic and changing, never static.

Conclusion

It is always prudent to appreciate and take note of the methods that successful scientists undergo into better furthering our understanding of the universe. Understanding successes and failures can enable a new generation of scientists to bring forth the world with ease. Albeit there may be many obstacles, surprises, or dangers on the path to enlightenment, persevering through them all will lead to the most saccharine of rewards; knowledge of the universe and all of its many mysteries.

References

1. The Discovery of Radium. Eve Curie (b. 1904). Pages 367-379 in Baigrie, Brian S. (ed.), Scientific Revolutions: Primary Texts in the History of Science. Pearson: Upper Saddle River, NJ, 2004.
2. Jones, Tommy Lee. "Quotes." IMDb. 1997. Accessed March 21, 2016. http://www.imdb.com/title/tt0119654/quotes.
3. Marie Curie Biography. Accessed March 22, 2016. http://www.biography.com/people/marie-curie-9263538.
4. Shadow Pictures. Wilhelm Conrad Röntgen (1845-1923). Pages 343-350 in Baigrie, Brian S. (ed.), Scientific Revolutions: Primary Texts in the History of Science. Pearson: Upper Saddle River, NJ, 2004.
5. "Wilhelm Conrad Röntgen - Biographical." Wilhelm Conrad Röntgen - Biographical. Accessed March 22, 2016. http://www.nobelprize.org/nobel_prizes/physics/laureates/1901/rontgen-bio.html.